SOLAR
CONSUMER
GUIDEBOOK

SOLAR CONSUMER GUIDEBOOK

Understanding the Process of the Solar Decision

MACKIE BEALL

Columbus, Ohio

The views and opinions expressed in this book are solely those of the author and do not reflect the views or opinions of Gatekeeper Press. Gatekeeper Press is not to be held responsible for and expressly disclaims responsibility of the content herein.

Solar Consumer Guidebook:
Understanding the Process of the Solar Decision

Published by Gatekeeper Press
2167 Stringtown Rd, Suite 109
Columbus, OH 43123-2989
www.GatekeeperPress.com

Copyright © 2022 by Mackie Beall

All rights reserved. Neither this book, nor any parts within it, may be sold or reproduced in any form or by any electronic or mechanical means, including information storage and retrieval systems, without permission in writing from the author. The only exception is by a reviewer, who may quote short excerpts in a review.

Library of Congress Control Number: 2022940818

ISBN (paperback): 9781662929533
eISBN: 9781662929540

This book is dedicated to Trevor Stille,
a brand leader to global markets.

It was his idea for me to share my solar knowledge
by writing this book, allowing consumers
to easily understand the buying journey.
Thank you, my friend.

Contents

1. Introduction: Advice from an Experienced Energy Consultant 1
2. The Process 11
3. Understanding Photovoltaics (PV) 15
4. Does Solar Increase Property Value? 16
5. Don't Be Pressured 18
6. Beware: Don't Be Sold 21
7. Financial Mindset/Decisions 24
8. Funding the Project 26
9. Selecting the Installer 28
10. Retirees 30
11. Evaluating the Proposal 32
12. Elements of the Sales Contract 35
13. Net Metering 36
14. Solar Array Design 38
15. Microinverters vs String Inverters 41
16. Solar Electricity Production 43
17. Panel Sizes 45
18. Solar Industry Terminology 46
19. Battery Storage 49
20. Warranties 51

INTRODUCTION
Advice from an Experienced Energy Consultant

The typical homeowner relies on an energy provider to supply electricity for their house and property. Each month, the energy provider supplies electricity for the homeowner to use. At the end of the month, the energy provider sends an invoice that may include use of the meter, franchise fee, on-peak charges, off-peak charges, and multiple taxes that are indiscernible unless you do research to understand terms like *DSIM* and *FAC* that are charged by kilowatt per hour. Oh yeah, the invoice also includes a charge for the amount of electricity consumed during the billing period.

Have you ever thought about the endless cycle of consuming electricity and paying each month for the amount consumed at the current rate charged by the provider? Does it concern you that electricity rates will never be cheaper than today's rates?

We, as well as our grandparents' parents have accepted the mindless process of continually renting our electricity. Solar energy allows the homeowner an option to take ownership of their energy production rather than renting electricity from month to month.

When speaking with someone about a serious topic, it is usually important to know where one's perspective is formed. I am a return-on-investment person. I work in the sustainable energy sector of the marketplace. My role is to help clients understand the impact of their decision to incorporate solar into their energy portfolio or maintain the status quo of renting their electricity and being hostage to future rate increases.

Not everyone will benefit from using solar energy to offset a percentage of electricity provided by your local electricity distributor. My recommendation is not necessarily to invest into a solar project for your property. My recommendation is for you to explore the possibility of taking ownership of your energy production.

The world of sustainable energy is driven mostly by "do-gooders" that seek to stop global warming and reduce greenhouse gases and the carbon footprint while making a profit for their business. I am not that guy. The financial component is my "why."

After a career spent providing goods for the athletic and home décor markets, I entered the sustainability world in 2009. As a private equity startup, I jumped into a new arena with both feet. My focus was providing solar film to glass surfaces to reject heat before entering a home. In 2017, the business was sold.

The world of solar was parallel to my solar window film business, so I maintained some level of awareness about solar energy and how it may impact consumers. It has not been until recent years that the investment was at least worthy of consideration. With tax credits, rising energy rates, blackouts and brownouts, and readily available funding sources, it is a new day for solar investors.

My goal for sharing information contained in this book is to help homeowners better understand how to approach the decision of whether to incorporate solar energy into their investment portfolio and take ownership of their energy production. If it meets your long-term investment expectations and is aesthetically pleasing, then you are a good candidate to join the solar community.

The process of evaluating solar energy likely starts when a person is made aware of the potential option to use solar for providing electricity. It may be an article that you read, an advertisement on social media, or awareness from a conversation. However solar energy entered your radar, the first steps are usually similar for every consumer.

Here is the common roadmap:

1. Do a "DuckDuckGo" search to identify local solar distributors. I suggest identifying two distributors to schedule an appointment. It is common that consumers invite three companies to earn their business. Because you are reading *Solar Consumer Guidebook*, there will not be a need for a third company to interview unless there is a large difference in the two proposals submitted initially.

2. After researching the distributors that initially appeal to you, contact the distributor to schedule an appointment to begin the process of evaluating their company and the solution offered to solve your problem.

3. An on-site or virtual meeting is recommended. When meeting in person or virtually, you have the added benefit of reading body language and providing the opportunity for you to measure your level of comfort with the solar consultant.

4. The solar consultant's goals are:

    a. **Build rapport** with you–the consultant will likely be eager to chat about anything in your home that will allow the consultant to make conversation about your interests. The technique is as old as the Model-T. Psychologically, the interaction is an "ice-breaker" intended to get you warmed up.

    b. **Evaluate** the project by asking questions to help the consultant understand the scope of the project. The questions must be asked, but your answers also provide the keys to knowing your hot buttons and how to 'sell' you. If the consultant is not asking probing and specific questions while eager to present a product and price, then you may not have the right solar consultant for your project.

c. A **proposal** will ultimately be presented to you by the solar consultant. The proposal serves as the main document that provides product manufacturer and number, product specifications, financial numbers including return-on-investment calculation, price per watt, and a legal section that provides the key terms that will be reflected in the forthcoming sales contract.

d. The **sales contract** is the final step in the sales process. Once agreement has been made on the amount of investment required and expectations set for delivering your project to completion are suitable with you, then there's nothing left but to review the sales contract.

e. **Sign the contract** to begin the post-sale process. Now that the agreement is in place, your communication with the solar distributor will become less interactive while your solar distributor orders your products, obtains proper permitting to install your solar array, and communicates with your electrical provider to exchange information. A lot of work is occurring while you're second-guessing the decision to change a lifelong habit of renting electricity while concerned if your investment truly will meet the return-on-investment figure that influenced your solar decision.

There are a couple of points that we'll address immediately to reduce stress related to setting an appointment to talk with someone about solar for their house. They are…

- All manufacturers of solar panels offer product sheets that contain engineering numbers and specifications needed to analyze the product.
- Most residential solar panels used in today's market are sized above 360w.
- Panels offered in America are mostly manufactured or assembled in America, Canada, or China.
- The production numbers associated for each solar panel should be reflected in the proposal provided by the solar distributor. The size of panel is only important because it takes more 360w panels to match the production of a 410w panel.
- Standard manufacturer warranties are 25 years for product, production, and labor.
- The manufacturer owns the warranty. Your warranty is backed 100% by the manufacturer. Furthermore, the manufacturer is required to have a third-party insurance carrier to back the

warranty in the event that the manufacturer's business fails.

Do you feel any better now? This type of information will be shared throughout the book. Product information is only one aspect of the solar decision experience. The company you select to deliver your project is also an important component in the process.

Personally, there are three key questions that would determine whether I would choose solar for my property. They are:

1. Do I have confidence in the solar company to deliver my project as quoted?

2. Do I have confidence in the manufacturer of the solar system?

3. What is the return on investment?

Aesthetics, Homeowner Association guidelines (if any), shading and size limitations are factors that the solar consultant should assist you with during the consultation.

Benefits of Solar

- Convert Variable Cost to Fixed Cost

    Even though a homeowner may pay a fixed monthly amount, the consumption and cost of energy are computed annually by the provider. The new monthly rate will reflect variances of usage and any changes in rates for projecting the new rate to be paid for the next twelve-month period. A fixed monthly rate from your provider is not a true fixed cost. It is a variable cost averaged over a twelve-month period. Solar energy is a fixed cost based on the amount of offset your solar array provides.

- Increases Home Value

    The national average is a $15,000 increase to your property value. I believe it is more meaningful to think in terms of a percentage. Depending on the area of America that your property is located, the amount of increase may vary greatly. Zillow states that 4% is a good number to use in estimating the value of solar for a residential property. There are other sources that estimate higher percentages.

- Tax Credits

    Many working Americans have an appetite for tax credits associated with a solar project. The United States Congress determines the percentage of tax credits in relation to the cost of the project that a solar investor may receive. Tax credits have been declining in recent years. The most popular types of incentives are:

    > Investment Tax Credit (ITC)

The federal government incentivizes homeowners when a renewable energy project such as solar is installed on a property. Congress has authorized a 26% tax credit for completed installations in 2022 and 2023. At this writing, there are no investment tax credits beyond December 31, 2023. It is wise to consult with a certified public accountant prior to signing a contract in order to verify that you will benefit from the ITC.

For example, your solar project is a $40,000 investment that will be completed prior to the expiration of the current ITC guidelines. Your calculation for IRS-Form 3468 will be $40,000 x 26% = $10,400 ITC. For most taxpayers, amount will qualify as a credit against your tax liability.

- Solar Renewable Energy Credits (SRECs)

    Some states offer the option for the homeowner to earn additional income by allowing electricity to be sold independently to the utility company. This exchange allows utility companies to meet state standards known as Renewable Portfolio Standards (RPS). The RPS regulation requires utilities to purchase a portion of their electricity from a renewable energy resource.

Electricity is the commodity that is sold as an SREC. Net metering credits may also be earned by the homeowner for excess solar energy production.

Supply and demand are critical factors that help in determining the price paid by the utility. A shortage in energy production will drive market prices higher. On the other hand, excess production will lead to lower prices paid to the homeowner. Market conditions in each state offering SRECs will fluctuate.

- Rebates

Some state and local governments offer rebates for new solar adaptors. Energy providers may also offer rebates to its customers.

- Reduces Electricity Usage

Offset is the term used to know how much electricity is being saved. Since solar is a long-term investment, many proposals offer solar savings over a twenty-to-twenty-five-year period.

- Environmental

Solar is an asset class that provides an eco-friendly solution for generating electricity. As a solar customer, you are reducing your personal environmental footprint by reducing the amount of carbon-emitting electricity from the grid.

The Process

Your responsibility is to be prepared by having questions about the company, the representative, the products, and funding the project. The solar consultant will ask for the past twelve months of consumption data. This is important to properly size the system.

Step 1: Inquiry Is Made by Prospect

As the homeowner, you are the prospect. The solar consultant will think of you in terms of becoming their new client.

The prospect typically makes an inquiry from a meeting that took place at an event (social circles, trade shows, etc.), social media, paid promotions, blogs, or a referral from one of the satisfied clients of the solar company that you will contact to schedule an appointment.

The initial phone conversation with a prospective solar distributor is the perfect time to gauge a potential fit with the solar consultant. If the consultant is asking good questions while making the conversation comfortable, then you are off to a good start. If the solar consultant is driving the conversation toward products and sales tactics, then at least you will know that the solar consultant is eager to make a sale.

Step 2: Meeting (On-Site or Virtual)

This is where the solar consultant attempts to make a personal connection with their prospect and a needs analysis is performed. This is also the time when a site survey (shading, roof condition, electrical panel) is conducted to verify that you are a good candidate for solar. The solar consultant will evaluate the amount of shading on your roof in order to know how to best design the solar array. There are several software packages for the solar company to choose from to maximize the quantity and positioning of the solar panels on the roof.

Step 3: Solar Array Design

Software used by the solar consultant will determine how many panels will be required to maximize your investment. The size of the system is determined by your past twelve months of consumption, the amount of desirable space available on the roof, and limitations dictated by your provider, homeowner association, or governmental agencies.

After performing the needs analysis, sizing of the system, and design of the solar array, the solar consultant is now ready to present their findings.

Presentation of the Project

The proposal is generated by the solar company. The proposal should show the placement of the solar panels, manufacturer, number of panels, size of panels (kW), return on investment, payback period, investment required to complete the project, monthly payment (if financing the project), and contract.

Q&A

This is the time when critical information is being processed by the homeowner. Good questions to ask:

- Tell me about the company.
- Why should I trust you to provide and install my solar system?
- What is the warranty?
- How long is the process of installing from the time I authorize you to move forward to the time the switch is flipped and my solar array is producing energy?
- If I sell my house, is the warranty transferrable?
- Is there a down payment required? If so, what are the payment terms?
- If I move forward, what is the process from start to finish?
- Will subcontractors be used for installation?
- Will I be able to monitor the performance of my solar system?
- In the event of hail damage, is that a homeowner insurance claim?

- Does solar increase the value of my house?
- Does solar cause my property tax to increase?
- What financing options are available?
- Do I receive rebates, tax credits, and incentives associated with the project?

Understanding Photovoltaics (PV)

Photovoltaics (PV) refers to the generation of electricity through sunlight by using solar panels as a source for renewable energy. Photoelectric cells harness the sun's rays.

An individual PV cell is small and typically produces 1 or 2 watts of power. The cells are often less than the thickness of four human hairs. The cells are protected by materials in a combination of glass and/or plastics.

The PV cells are connected in chains to form larger units known as modules, or panels. Modules can be used individually, or several may be connected to form arrays. The arrays are connected to the electrical grid as part of a complete PV system.

Does Solar Increase Property Value?

It depends on whom you ask. I have spoken with at least one realtor who has the opinion that solar energy provided to a residence does not increase property value. Most do not agree with this stance.

The Office of Energy Efficiency and Renewable Energy (EERE) states, "Solar panels are viewed as upgrades, just like a renovated kitchen or a finished basement, and home buyers across the country have been willing to pay a premium of about $15,000 for a home with an average-sized solar array."

Zillow, along with a study from the National Bureau of Economics Research, states that solar on a residential property increases the home value by 4% compared to homes without solar.

A study by Lawrence Berkley National Laboratory uses the formula of a $5,000 resale value increase for every kilowatt of solar installed.

Regardless of the valuation method of solar for your property, it is common sense that some buyers will gravitate to a property using solar energy to offset their consumption demand over a property relying 100% on their energy provider. There will also be buyers who will not be inclined to value solar as an asset to the property.

If you seek an advantage in a competitive market, then solar will likely make sense.

Also consider that solar panels on your property w ll not increase Personal Property Tax liability.

Don't Be Pressured

One of the solar sales models used by multiple companies is designed to get you to commit to purchasing solar during the first meeting with the solar consultant.

For the average residential solar project that I install, the customer is investing approximately $40,000. Do you need to make an investment decision that involves tens of thousands of dollars while meeting with a solar consultant the first time? Is there a time deadline that requires you to sign a contract after spending an hour with the consultant at your dining room table? If not, it is wise to mentally digest the details of your meeting and evaluate the numbers provided in the proposal given to you by the solar consultant.

It should serve as a warning to you if the consultant attempts to pressure you into committing to anything other than a follow-up phone call or meeting. Take your time until you feel a level of excitement and confidence about moving forward with a solar project.

To the solar consultant, you are one prospect in their "pipeline." The solar consultant's role is to provide information for you, address questions, communicate, be truthful, provide customer service, and generate revenue for their company.

You'll know if you have a "salesperson" if you are required to endure a PowerPoint presentation. The Solar

Consultant will use the sales meeting as a process of creating rapport, answering questions, reviewing your electricity usage, conducting an audit (determine roof condition, shading, electrical panel evaluation), 'pitching' (a sales technique that attempts to persuade someone to purchase), telling compelling stories, and asking for the order.

If the consultant and solar company are ethical, there is no issue with this approach. It's the same technique that has been used in the school of sales for decades. However, beware of the solar consultant who fails to ask questions that allow you the opportunity to share your thoughts and goals for your solar project.

I am not an advocate of having several solar companies meet with you to provide pricing for your consideration. Two of the most reputable resources available to the homeowner are *EnergySage.com* and *SolarReviews.com*.

My recommendation is to research solar distributors and identify all that present the most compelling reasons to earn your business. Take the time to read customer reviews regarding their experience with the prospective solar provider. What percentage of reviews are positive?

For those customers that did not have a satisfactory experience, how did the solar provider respond to the issue? Was the issue resolved to the satisfaction of the homeowner?

From that point, select two companies to schedule an in-home meeting. If the solar consultant for both companies is someone you trust to deliver your project, then there's likely no reason to continue the search.

Pricing for both proposals should be within a couple of percentage points if both proposals have sized the project correctly. If there is a large discrepancy, then we're likely not comparing "apples to apples." If this is the case, you may choose to invite a third company to the table to help determine the variance. If you're comfortable asking probing questions, then you may choose to address your questions to each solar consultant until you determine the reason for the variance.

Beware: Don't Be Sold

The typical sales process begins with "Hello." At that time, the salesperson will attempt to build rapport with you. The conversation may begin with current events or perhaps an object sitting on a shelf or table that may indicate a particular interest that you may have, such as golfing, hiking, church involvement, or any other personal interest.

Once "small talk" has taken place, the salesperson may ask for the conversation to continue at your kitchen table. You will be asked to provide your electric bills for the past twelve months. At that time, the salesperson may begin talking about services offered beyond solar. One common service may be referred to as an "Energy Audit." An energy audit will cover things like attic insulation, LED lighting, hot water heater blanket, etc.

Now the pitch begins. The salesperson may use a computer or laminated spiral binder to walk you through points of information designed to bring your "pain" to the surface. Visual depictions of attic insulation, R-values, attic ventilation, air duct seals, radiant heat gain, water heater, LED lighting, refrigerator, thermostat, average home usage, rising energy rates, tax credits, and rebates.

After the sales pitch, now comes the solar sales part of the meeting. The salesperson will explain the benefits of solar and how solar works. They may tell you something about the history and growth of solar.

Monocrystalline solar panels are "old school." Monocrystalline panels are now associated with products that are sold with deep discounts with many manufactured in foreign countries.

Polycrystalline solar panels are more eco-friendly and efficient than the older-generation monocrystalline panels. In today's market, the solar consultant may offer both as an option for the price-conscious prospect.

Eventually, the salesperson will explain the difference in polycrystalline and monocrystalline panels and how wiring configures your system into the electrical panel. Of course, a good salesperson will not miss the opportunity to tell you how their panels are the best ever manufactured.

The last step in the sales pitch may be sharing some points of interest about the company that will provide products and install your project. Throughout this meeting, the talking points are all designed to make your brain focus on the pain that you are feeling from rising energy costs and building value so that the investment amount associated with your project doesn't seem like a big number.

Price is most often the biggest concern that the salesperson must overcome. Price is also the topic that provides the biggest reason to take the step to solar. A good salesperson will say that you are currently paying 100% interest for electricity with no return on your investment.

If a salesperson attempts to pressure you to make a purchase decision at your first meeting, beware! Why should anyone feel pressure to make a $40,000 decision within an hour of meeting with a salesperson? What I have found is that the company that seeks a "one-call close" is a company that offers less value than a company offering solar that is willing to allow the buying cycling to extend as long as the prospect needs to make a good decision.

Financial Mindset/ Decisions

You may not realize it, but you have a contract with your current electrical provider. The deal is that the provider sends electricity for your consumption. At the end of the month, you receive an invoice. If the invoice is paid, the provider will continue to provide electricity for your consumption. If the invoice is not paid, then services will be interrupted and additional fees will be required to reinstate service.

You are renting electricity from the provider. When a consumer makes the decision to take ownership of their energy production by installing solar on their property, a variable expense is converted to a fixed cost.

Many people who are approaching retirement and seeking to reduce operating expenses often consider solar as a tool to save money and help for monthly budgeting purposes. When the investment is made into solar energy, you are locking in today's cost of energy and guarding against future increases in energy cost.

The cost of energy is rising. Rate increases vary by provider and the region in which you reside. It is wise to research the historical data from your provider while paying attention to current events to project future changes in your rate.

Solar does not cost. Renting electricity from your

provider is a cost. Solar is an investment into your energy portfolio. There is a return on investment that is calculated in relation to your cost of energy, amount of consumption, and the amount of investment required to func your project.

Funding the Project

The consumer has four primary options to fund their solar project. They are:

1. Cash

This may be from your bank account, the sale of an asset, converting an investment into cash, or you securing funding from another source. If you bring funding to the table, you will be considered a cash customer. There may or may not be any advantage to paying cash for your project. That question will be answered by your solar consultant.

2. Third-Party Financing

There are multiple companies that offer financing for your solar project. There are dozens of funding options available. Some are low interest percentage with substantial loan origination fees attached while some offer higher interest rates with small or no loan origination fees. Your solar consultant should be versed on a loan package that suits your best interest.

3. Lease

This option is most attractive for anyone who wants no out-of-pocket expense and no risk in the production and maintenance of their solar system. The lease is also for anyone who does not benefit from the ITC and doesn't consume a lot of electricity. A lease will require a credit-worthy buyer when the time comes to sell the house.

4. Power Purchase Agreement (PPA)

Note: PPAs are not available in all states.

A contractual agreement between a third-party solar developer and a homeowner. The homeowner has no upfront costs and is not responsible for maintenance on the system. The homeowner should expect to pay a fixed price for electricity, and the contract will likely include an escalator to account for degradation and inflation-related costs. The solar service developer is responsible for financing, installation, and maintenance over the life of the contract. The developer also receives any tax credits and incentives related to the project.

Selecting the Installer

In choosing to go solar, you are preparing for a long-term investment that typically involves a $20,000–$100,000 commitment. How do you go about the process of identifying the best person to deliver your project? Here are five areas to research:

1. **Longevity**

    The world of solar is somewhat akin to the "Wild Wild West." The life cycle of a business taught on college campuses doesn't necessarily apply in today's solar industry. A solar company in business for five or more years is considered a mature operation.

2. **Reviews**

    How many clients have taken the time to provide feedback of their experience with the installer whom you are considering to deliver your project? What percentage of reviews are positive and negative?

3. **The Staff**

    Ask which specific staff members (and their title) will be involved in your project from start to finish. The response from your solar consultant will be an indicator of understanding their organizational structure. For instance, the site survey may be performed by one person while the operations manager has a different set of responsibilities. It may be a warning sign if one person is responsible for multiple activities during the procurement, installation, and post-sale stages of your project.

4. **Organizational Involvement**

   Membership into recognized industry-specific organizations is one sign of the level of professionalism exercised by the installer. SEI (Solar Energy International) is one organization that is known for the level of training provided to its members. Another organization that indicates a good association for the installer is NABCEP (North American Board of Certified Energy Providers). There are also various statewide and regional organizations. Beyond membership, if the installer is active in leadership roles, then the company is more likely to exercise professional business practices.

5. **Warranties**

   The manufacturer owns the warranty. The warranty is backed by a third-party insurance policy to ensure that customers are secure in the event that the manufacturer goes out of business or their financial statement reflects the inability to service warranties.

Retirees

Does solar make sense for someone entering or currently in retirement? The answer may be yes or no.

If anyone seeking an average annual return on investment in the 6%–12% range for the next twenty years, then solar is worth strongly considering. Factors that determine the exact return on investment are cost of electricity, how much you consume each year, and in which region of America you reside.

Offset is an important term to understand. *Offset* simply means the percentage of electricity your solar array will provide and the percentage of electricity your utility provider will contribute in meeting your average monthly energy demand.

If you are someone approaching retirement and seeking to reduce cost-of-living expenses, then solar will likely be an attractive investment. If your offset is 90%, then you are converting 90% of your energy budget to a fixed cost. The remaining 10% paid to your energy provider will remain subject to rate increases in the future. The good thing in this example is that a 6% rate increase has a substantially smaller effect on someone with a 90% offset.

Example:

Your average monthly electricity bill is $100. You have solar with a 90% offset. This means you pay your energy provider $10 per month. Your provider has a 6% rate increase. The 6% increase is calculated on the $10 rather than $100. Therefore, your new bill will increase $0.60 monthly as opposed to $6.00.

For someone in retirement, the answer may not be quite so clear. Keeping in mind that the payback for solar is typically in the eight-to-twelve-year range, the question will pertain to the future owner of your house. If the house is to be willed to a family member or sold on open market, then solar will likely be attractive. The family member will benefit without question. If the house is to be sold on the open market, solar will likely attract a certain buyer because of the solar powerplant.

Evaluating the Proposal

There are several key metrics to focus on when reviewing the solar proposal. Don't become focused on the number of panels or the price tag related to your project. Your focus should be on price per watt, the average annual utility escalator (stated in terms of a percentage), and average annual return on investment.

Price Per Watt (PPW)

Price per watt (PPW) is the key number for solar when comparing proposals. Sometimes it is confusing to view two proposals that are slightly different in sizing with different manufacturers and consumption data. The PPW is the equalizer. The lower the PPW, the more attractive the project. PPW varies from region to region.

Panels

Solar PV panels are typically not exclusive to any dealer. Solar dealers commonly use more than one manufacturer. Most dealers use this approach because they want access to the latest version of solar panels.

Microinverter

For solar PV mounted on the roof, this is an important component to consider. The microinverter is paired with the solar panel. The microinverter collects direct current (DC) output from the sun and converts it to alternating current (AC). The transfer of energy from DC to AC both regulates and optimizes energy flow used by your house and fed back to the energy grid. Microinverters also allow the manufacturer and homeowner to monitor the production and efficiency of your system.

The proposal should clearly state the manufacturer and model used for your project.

Design

Make sure that your solar array is not adversely affected by shading. If during the site survey it is determined that a change in design is required, then a new contract or addendum should accurately display any changes in the financial and production data. If a design change is required, that is not necessarily the fault of the solar consultant. Software design companies are sometimes so close that solar panels may exceed local code standards and require removing at least one panel.

Escalator

The *escalator* is the average annual percentage rate increase used to determine future savings created from your solar system. The percentage will vary depending on the history of rate increases by your current provider. If your proposal uses a 4.5% escalator and the actual increases average 3.4%, then there will be a sizable negative difference in your return-on-investment calculation.

Consumption

Make sure that your proposal accurately calculates and uses your annual consumption for your proposal. If the companies providing your proposal use a different annual kilowatt-hours (kWh), then the financial and production data will be skewed. Let's assume, your actual annual energy consumption is 20,000 kWh and your proposal uses 16,000 kWh. Your offset and return on investment will appear more attractive on the proposal. It may take several months before you realize that your expectations are not being met. This scenario may be avoided by knowing to verify that your actual annual consumption matches the amount used on your proposal.

Elements of the Sales Contract

When presented with a proposal, the client should expect to see key elements of a contract clearly stated. The topics essential to an effective proposal are:

1. Statement of date, customer name, and address of the proposed project
2. Terms and Conditions
    a. Products and Components
    b. Scope of Work
    c. Price and Payment
    d. Timeframe
    e. Warranty Information
    f. Legal Section
3. Pricing
4. Payment Terms
5. Project Components
6. Warranties
7. Insurance Coverage Limits for Installer
8. Permitting Responsibility
9. Timeline for the Project

Net Metering

The last step in the installation process is changing your meter to a net meter. Your energy provider will install the new meter after completion of the solar installation. Once inspection of the completed project is performed by the energy provider, then the installation of the net meter is scheduled.

The net meter allows overproduction of your solar system to send excess energy back to the grid. Yes, the meter runs backward when your solar array is producing more electricity than you are consuming.

Most energy is produced during the middle part of the day. This is a time that most persons are not at home and consuming electricity. The net meter now runs backward and sends the excess energy back to the grid for consumption at a later time.

Do energy credits "roll over" from month to month? It depends on the utility provider. It is not uncommon that credits forward to the next month until the end of the year. At end of the year, there is a date referred to as "True-Up Day." This is the day that your account is reconciled. If there is an energy credit on your account, then one of two things will happen. Either you lose the credit and your account starts "clean" for the next twelve months, or your provider will send a bank check paying you for providing energy to the grid. Sounds good, doesn't it? Not so fast, my friend.

Your energy provider will likely refund your energy production at approximately 25% of what you pay for the same energy. This greatly discounted rate spotlights the importance of accurately sizing your solar system.

The idea that your energy provider will purchase from you any excess energy that is produced is appealing for many people. On the surface, it sounds like an interesting concept. However, purchasing an oversized solar system to sell excess electricity to your energy provider at a greatly reduced rate has a significant negative impact on your return on investment.

Solar Array Design

Software is commonly used to calculate the proper sizing for a solar system and placement of the panels on your roof or ground to provide maximum efficiency of energy production.

Sizing means placing each solar panel on your roof at the optimum angle to the sun and having the exact number of solar panels to create an attractive return on investment.

Southern exposure offers the most production for net metering purposes. Southwest exposure also offers high efficiencies. East and west panel placement are also used for solar design.

Output for panels located on the east and west roof are typically reduced by approximately 15% when compared to south-facing panels. North-facing panels are 30% less efficient than south-facing panels.

Latitude and the pitch of your roof will impact solar production. Depending on the pitch of your roof, the racking system used to install your system will be set at the proper angle to maximize production.

Regardless of whether you have south roof available, solar may still make sense for you.

Compensation may be made by simply adding more panels. The return on investment will be negatively affected but may still meet your investment goals.

For the solar consultant, the goal is to have a system designed for maximum efficiency.

Fortunately, modern solar is supported by software that takes the human error factor out of the equation.

There are factors to consider before, during, and after your solar installation. Energy production may be lost by:

- **Debris**–make sure that leaves, plastic bags, and other factors do not collect under or on top of the solar panels.

- **Snow**–considering that solar panels are approximately 10 degrees warmer than the environment, small amounts of snowfall may quickly dissipate in less extreme temperatures. When multiple inches of snowfall occur, your solar panels may not produce for several days.

- **Wiring**–inspections are made by the installer, utility provider, and possibly city code official to verify that the system is ready to activate once the net meter is installed. However, once the system is producing energy, then there is the possibility of critters (squirrels in particular) damaging wiring underneath the panels. There are products available to combat the furry enemies of solar.

- **Shade**–this factor should be considered and discussed early in the sales cycle. Sometimes tree limbs may be removed to accommodate solar. Other times, trees may prevent solar from being a viable option for the homeowner.

- **Age**—as years pass, your solar system will incrementally degrade in production. *Degradation* is the term used to describe the amount of efficiency lost each year on each panel that is manufactured. The annual degradation is stated on the product specification sheet. The proposal software takes degradation into account when projecting long-term production and financial calculations.

Microinverters vs String Inverters

The inverter is an electronic device that serves as the "brains" for a solar system.

The inverter changes the DC electricity produced by solar panels into the AC electricity used to power your house.

The inverter also relays information to you, usually through your smartphone, about how well your system is producing energy for your house.

There are three types of inverters offered on the market. They are:

- **String Inverters**—sends the power produced by solar panels to a central inverter that converts DC to AC power and then sends usable AC electricity for use in your home. String inverters connect 8-15 panels and are often used for ground mount systems. Some string inverters do not provide consumption data (how much energy you're using), only production data (how much your panels are producing).

- **Microinverters**—converts DC to AC at each individual panel and allows for monitoring and power regulation at the panel level. Microinverters are used primarily for rooftop systems because of shading factors.

- **String Inverters + Power Optimizers**—this combination uses both string and microinverters to maximize each individual solar panel output. There is usually one DC optimizer per panel.

- **Hybrid Inverters**—combines solar inverter technology with battery inverter capability.

Solar Electricity Production

There are five factors that should be considered for any solar project. They are:

1. **Shade**

    If your roof is heavily shaded, solar may not be the best option for you unless you are willing to trim or remove trees to provide direct sunlight to your roof.

2. **Seasonality**

    Focus on the twelve-month period when evaluating solar. In many parts of America, December through February will provide significantly less energy production than June through August. The solar consultant should be forthcoming with information to help you understand production that is projected for each individual month of the year.

3. **Available roof space**

    Some roofs may not have enough available space to meet the solar requirements for an attractive return on investment.

4. **Tilt**

    The direction your roof is facing, its location, and even your roof's pitch have significant impact on solar efficiency. Ideally, solar panels should be at the same angle as the latitude where they're mounted. Pitches between 30 degrees to 45 degrees usually work well in most scenarios.

5. **Azimuth**

    The solar azimuth angle is the compass direction from where the sunlight is coming. At noon, the sun's light comes from the south in the Northern Hemisphere and from the north in the Southern Hemisphere. The placement of your solar panel may greatly impact the production of your solar array. The wrong azimuth angle could reduce the energy output of an array by >30%.

Panel Sizes

The most popular size panel is 60 cells. However, there are many size variants that are manufactured for you to purchase the solar panel that most efficiently fits your consumption demand. The 72-cell and 96-cell panels are becoming more popular.

Panels sizes:

- 60-cell panel   39.0"w × 66.0"h × 1.3" to 1.6"d
- 72-cell panel   39.9"w × 77.0"h × 1.3" to 1.6"d
- 96-cell panel   41.5"w × 62.6"h × 1.38"d

Solar Industry Terminology

Annual Solar Savings—the amount of money saved after installation of solar relative to a nonsolar building.

Array—an interconnected system of PV modules that function as an electricity-producing unit.

Battery Storage—a device that transfers energy from electric to chemical form and vice versa. During discharge, chemical energy is converted to electric energy and is consumed in an external apparatus.

Converter—a unit that converts a direct current (DC) voltage to another DC voltage.

Degradation—the annual declining percentage of a solar panel's efficiency. Degradation is caused by exposure to the elements and normal wear on the solar panel.

Electrical Grid—an integrated system of electrical distribution covering a large area.

Grid-tied Solar—a PV system connected to the energy grid without a battery storage requirement. Grid-tied systems make use of net metering and generating credits during peak production months by using grid-tie inverters to communicate with the energy grid. The main disadvantage is their inability to provide power when the grid is not functioning.

Hybrid System—a PV system that includes other sources of electricity generation, such as wind or diesel generators.

Inverter—a device that turns the DC electricity into AC electricity after solar radiation is transformed into DC electricity from the solar panel.

Interconnection—the physical connection between the electrical grid and your solar array.

Kilowattt (kW)—a standard unit of electrical power equal to 1,000 watts.

Kilowatt-Hours (kWh)—a way of measuring how much electricity is being consumed. A kWh equals the amount of energy you would use by keeping a 1,000-watt appliance running for one hour.

Net Metering—a credit that the homeowner receives from the power provider for returning power to the grid for use at a later time.

Off-grid—a property that is not connected to a power grid.

Offset—the percentage of electricity that your solar system provides. The formula is the amount of electricity your solar system produced in a twelve-month period divided by the total amount of electricity used during that same twelve-month period.

Photovoltaic (PV)—a device that generates power by absorbing energy from sunlight and converting it into electrical energy through semiconducting materials (solar modules).

Power Purchase Agreement—a contract between the homeowner and a solar developer. The homeowner receives electricity at a stated price without risks associated with a solar system. The developer receives any federal and state incentives associated with the project.

Shading—the amount of light that is reduced from direct contact from the sun.

Time of Use (TOU) Billing—the time of day that utility companies tend to charge more for electricity. TOU is usually measured in four-to-six-hour time periods. For example, one utility provider may have peak demand between 4:00-8:00 p.m. while another provider may be 2:00-5:00 p.m. Rates during peak demand periods are typically significantly higher than off-peak rates.

Battery Storage

Solar is an investment. Battery storage is a cost. Some companies push consumers toward battery storage. Until a needs analysis has taken place, be leery of a solar consultant who is eager to promote storage at this point in time.

Battery storage units provide power to your property regardless of the time of day or weather without relying on the energy grid. Even if your grid service is uninterrupted, the battery system may be programmed to use energy stored in its cells rather than using the utility for power.

Do you NEED battery storage? There are two primary questions to be answered to know if you need storage. Do you have frequent and extended outages? Do you have medical or other equipment that requires 24/7 electricity?

If the answer to either of these questions is yes, then you may easily justify the investment decision to incorporate battery storage into your solar design.

Why don't all solar projects include battery storage? The answer is simple—COST. At this time, battery storage is considered expensive. Storage should not be considered for return-on-investment decisions. Storage should be considered more so as insurance against outages and interruptions to daily living.

Compared to the timeline of solar development, battery storage is currently in the stage where manufacturers are racing to provide the latest/greatest battery unit to the consumer. Pricing for battery storage is high. For example, a single American-made battery will have a net cost over $14,000. Additional batteries will cost approximately 15% less when attached to the same project.

Low-voltage items such as TV, lights, refrigerator, computer monitor, the garage door, and ceiling fans may be operated by one battery unit. By being mindful and efficient with usage, a single battery may store enough energy for a few days of usage.

A "whole home solution" refers to the ability to operate your household at approximately 70% of normal activities. This means that the HVAC system may be used along with other high-voltage items (clothes dryer, space heater, hair dryer, EV charging station). This solution will require more than one battery.

It is important to thoroughly understand the details of what you should expect by incorporating battery storage into your energy equation. The solar consultant should be knowledgeable and have resources to accurately calculate that your expectations will be met when you decide to use storage in your energy solution.

If you cannot justify battery storage at this time, then you are seemingly in a good position. Be patient and allow time to pass so that technological advancements and economies of scale occur before adding storage. Also, keep in mind that a gas-powered generator may be an option.

Warranties

It is standard practice in the solar industry to offer a 25-year product, production, and labor warranty. Some inverter manufacturers offer shorter warranty periods with an option to purchase an extended 25-year warranty.

The warranty is backed by the manufacturer, not the installing company. The manufacturer is required to have a third-party insurance carrier to make sure that the company can service its obligation in the event of a failure in their product.

Provided that you have chosen a solar product manufacturer that offers a quality product, the likelihood of ever needing the warranty is small.

www.ingramcontent.com/pod-product-compliance
Lightning Source LLC
LaVergne TN
LVHW011859060526
838200LV00054B/4420